INVESTIGATING
NUMBERS

Ed Catherall

Wayland

Investigations

First published in 1982 by Wayland Publishers Limited
49 Lansdowne Place, Hove, East Sussex BN3 1HF, England
© Copyright 1982 Wayland Publishers Limited
ISBN 0 85340 990 0

Illustrated and designed by David Anstey
Typeset by Tunbridge Wells Typesetting Services Ltd
Printed in Italy by G. Canale & C.S.p.A., Turin

Contents

Chapter 1 The first number systems

Numbers, numbers everywhere

How old are you? What is your birthday date? What is your home address?
How many numbers are there in that address?
What telephone numbers can you remember?
List the emergency numbers for your area.

Ask your friends if they have lucky and unlucky numbers. Make a list of these numbers.
What do you notice?

How many different things are numbered on your street?
List the different things that are numbered which you pass on one journey to school.

Look at the games that you play.
How many of these games have numbers?
How many games use dice for numbers?

Look in magazines for pictures that have numbers on them. Cut out these pictures and mount them on a large sheet of paper to make a number mural.

The language of numbers

Numbers are codes invented by humans.
Count to 10. Each number has a
name that was invented.
The name or sign of a number is its
numeral.

Can you count in a foreign language?
Can you spell these foreign numbers?

Look in foreign language dictionaries.
Write the words for 1, 2, 3, 4, 5, 6, 7,
8, 9 and 10 in three different languages.

The ancient Chinese had a number
code. (Picture 1)
Chinese numerals were written
downwards.
Write your age in Chinese numerals.
What is your school's telephone
number in Chinese numerals?

From 11 to 19, the Chinese
added. (Picture 2) After 19
they multiplied and added. (Picture 3)
Write some numbers in Chinese.

Chinese numerals

① 一 —1　　七 —7
二 —2　　入 —8
三 —3　　九 —9
四 —4　　十 —10
五 —5　　百 —100
六 —6　　千 —1000

② 圡 —11
圭 —13
圥 —17

③ 21— 卅　 2 × 10
+1

675— 六百七十五　 6 × 100
7 × 10
+5

5

Ancient number systems

Over 5,000 years ago there were two number systems in the Middle East. The people of Sumer in southern Babylon wrote their numerals on soft clay.

Each wedge shape, ∨ , counted as one. For 10 the shape was turned ◄. They continued counting until they reached 60 which was a large wedge ▼.

The Sumerians counted in 60s. We count in 10s, but have 60 seconds in a minute and 60 minutes in an hour and 360° in a circle.

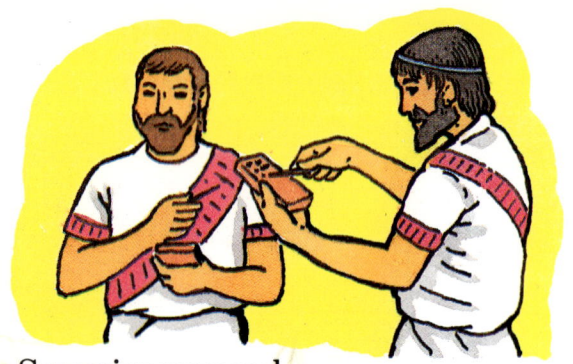

Sumerian numerals

∨ —1   ∨∨∨∨ / ∨∨∨ —7
∨∨ —2   ∨∨∨∨ / ∨∨∨∨ —8
∨∨∨ —3   ∨∨∨∨∨ / ∨∨∨∨ —9
∨∨∨∨ —4   ◄ —10
∨∨∨ / ∨∨ —5   ◄∨ —11
∨∨∨ / ∨∨∨ —6   ▼ —60

◄◄◄∨∨∨ / ◄◄∨∨∨ —56

The Egyptians, 5,000 years ago, wrote their numerals in ink on papyrus.

Each single stroke, ∕ , counted as one. For 10 the numeral was ∩ . The Egyptians counted until they reached 99 and then made a new numeral, 𝟫, for 100.

Write numbers in Sumerian and Egyptian.
Which system is easier to use?
Which system is most like our system?
Find out all you can about Sumerians and ancient Egyptians.

Egyptian numerals

∕ —1   ⦀/⦀ —9
∥ —2   ∩ —10
⦀ —3   ∩∕ —11
∥∥ —4   𝟫 —100
⦀/∥ —5   𝔤 —1000
⦀/⦀ —6   ⌒ —10,000
⦀⦀/⦀ —7   𝒪 —100,000
⦀⦀/⦀⦀ —8

𝔤 𝟫𝟫𝟫/𝟫𝟫𝟫/𝟫𝟫𝟫 ∩∩∩/∩∩∩ —1990

6

The Greek and Mayan number systems

Over 2,500 years ago the Greeks used the number system shown opposite.
Use this Greek system to write your age and the ages of members of your family.

Write the numbers represented by 1B, 1F, KA, MF, and NZ.
Write the Greek numbers for 32, 49, 61, 84, 99, 284, 569, 728 and 833.

Try doing some simple arithmetic problems using this system. Why is it difficult to use?

The Mayans of Central America used a simple system based on only three symbols.
They counted using their fingers and toes, so their system was based on 20.
Write numbers using the Mayan system.

Greek numerals

A	—1	Ξ	—60
B	—2	O	—70
Γ	—3	Π	—80
Δ	—4	ϙ	—90
E	—5	ϟ	—100
F	—6	Σ	—200
Z	—7	T	—300
H	—8	Y	—400
Θ	—9	Φ	—500
I	—10	X	—600
K	—20	Ψ	—700
Λ	—30	Ω	—800
M	—40	ϡ	—900
N	—50		

Mayan numerals

•	—1	••• (bar)	—8
••	—2	•••• (bar)	—9
•••	—3	(two bars)	—10
••••	—4	• over two bars	—11
—	—5	(shell)	—20
• (bar)	—6	(bars/shell)	—200
•• (bar)	—7		

Roman numerals

Look for Roman numerals. You can find them on clocks, watches and sundials.
Sometimes the date is written in Roman numerals.

Roman numerals were taken from the Roman alphabet — I, V, X, L, C, D, M. The Roman word for 1000 is mille and the numeral is M. We still use these in the metric system today.

Notice that smaller numerals to the right are added as in VI which is 6, VIII which is 8, and LXX which is 70.
Smaller numerals to the left are subtracted as in IV which is 4, IX which is 9, and CM which is 900. What are the numbers XVI, LV, and XCV?

Write this year's date in Roman numerals.
Write 18, 43, 49, 54 and 99 in Roman numerals. Now try adding these numbers in Roman numerals. Try doing subtraction, multiplication and division in Roman numerals. Check your answers by using our number system.

Roman numerals

I—1	XXV—25
II—2	XL—40
III—3	XLV—45
IV—4	L—50
V—5	LI—51
VI—6	LX—60
VII—7	XC—90
VIII—8	C—100
IX—9	CD—400
X—10	D—500
XI—11	CM—900
XII—12	M—1000
XIX—19	
XX—20	

MCMLXXXVIII—1988

Our number system

Over 2000 years ago the people of India developed a system with a different symbol for each number up to 9.

They also invented a zero.

Notice that none of the other systems has a zero (see pages 5, 6, 7 and 8).

By using a zero it is possible to write all the whole numbers with just ten symbols:

0 1 2 3 4 5 6 7 8 9

The Arabs learned this system from the Indians and gradually changed the form of their numerals.

With small changes these numerals are the ones we use today. We call them Arabic numerals because they look like the ones used by the Arabs, but remember it was the Indians who invented them first.

Indian numerals

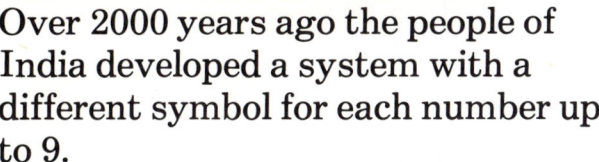

O —0	੫ —5
੧ —1	੬ —6
੨ —2	੭ —7
੩ —3	Z —8
੪ —4	ੲ —9

Arabic numerals

O —0	੫ —5
I —1	Ⴑ —6
੨ —2	7 —7
੩ —3	8 —8
੪ —4	9 —9

Place value

In order to write all the whole numbers using only ten symbols everyone must learn the place value code.
Write the number 1 and the number 10.
Notice that by adding a zero at the end of the number you increase that number ten times. (Picture 1)
By naming the columns you make a counting board. (Picture 2)

Use a ruler to draw a counting board on cardboard. (Picture 3)
Rule four columns on your board for units, tens, hundreds and thousands. (Picture 3)
Make counters by drawing around a coin on cardboard.
Cut out your counters.
Arrange your counters on the board and record the number that you make. (Picture 4)
Try counters in many different ways.
Always record the number.

How many different numbers can you make on the board using only two counters?
How many different numbers can you make using three counters, then four counters?
Can your friend make more numbers than you?

①

			1	One
		1	0	Ten
	1	0	0	One hundred
1	0	0	0	One thousand

②

Thousands	Hundreds	Tens	Units	
			1	One
		1	0	Ten
	1	0	0	One hundred
1	0	0	0	One thousand

③ A counting board

Thousands	Hundreds	Tens	Units	
			Total	

④

Thousands	Hundreds	Tens	Units
		●	
		●	●
●		●	●
1	0	3	2

Thousands	Hundreds	Tens	Units
●			
●		●	
2	0	1	0

10

Making an abacus

Thousands of years ago the ancient Egyptians and Chinese used an abacus for counting.
The Mexican Indians were using abacuses when Columbus came to America.

Find four identical knitting needles.
Find nine wine bottle corks.
Ask an adult to help you cut each cork into four discs. (Picture 1)
Use a nail to make a hole in the middle of each cork disc.
Push nine cork discs onto each knitting needle.

Find a thick piece of balsa wood.
Push the point of each knitting needle into the block of balsa wood. (Picture 2)
You have made an abacus.

Notice that each knitting needle represents a column in the place value code (see page 10).

Make the numbers 12, 22, 202, 2, and 121 on your abacus. (Picture 3)
Use your abacus to do addition and subtraction.

You can make a stronger abacus by making a balsa wood frame.
Remove the cork discs from each needle. Push the needles through the balsa wood base. Replace the discs and cut balsa wood to fit the other three sides. (Picture 4)

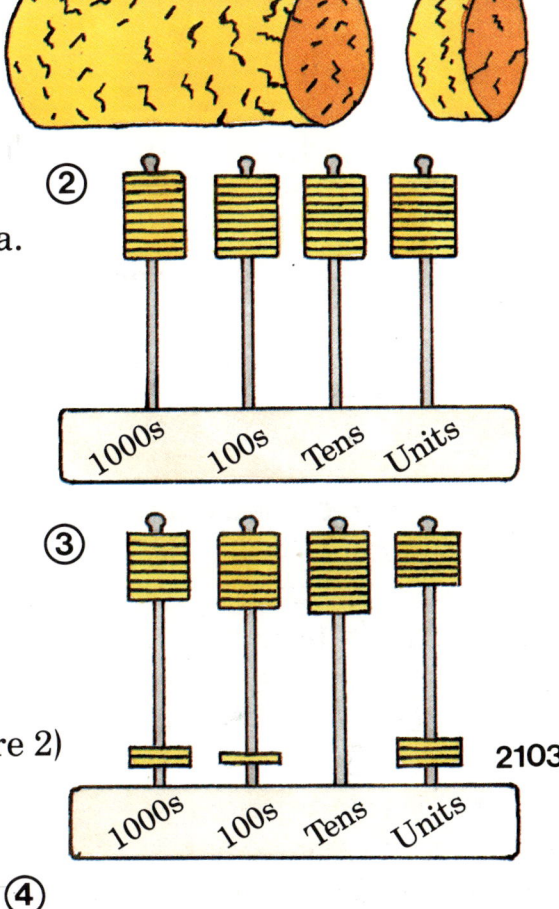

①

② 1000s 100s Tens Units

③ 1000s 100s Tens Units 2103

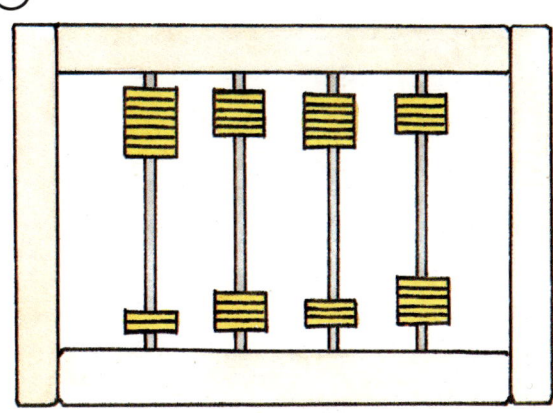

④

11

Chapter 2 Rules for numbers

Addition on a number line

Use a ruler to draw a long straight line.
Mark numbers on this line at regular
intervals in order from zero onwards. (Picture 1)
Move from 0 to 5. How many spaces
do you pass through? Mark this
move as a jump on your number line. (Picture 1)
Now move from 5 to 9. How many
spaces do you pass through? Mark this as
a jump on your number line. (Picture 1)
We can write your moves like this:

5 spaces + 4 spaces = 9 spaces
$5 + 4 = 9$

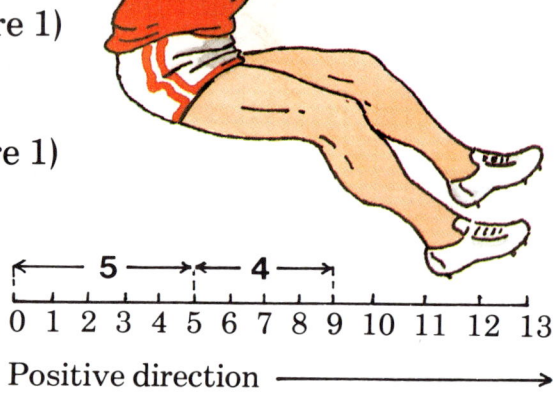

①
Positive direction ⟶

Use your number line to show the jumps —
4 + 5; 6 + 3; 2 + 7; 1 + 8; 9 + 0.
What do you notice?
When you jump to the right you are
moving in a positive direction.

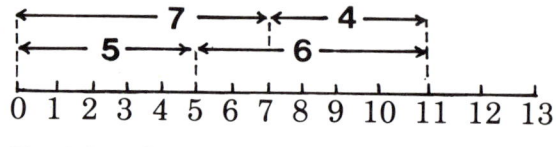

②

Positive direction ⟶

$$7 + 4 = 11$$
$$5 + 6 = 11$$

③ Addition square

Use your number line to record
jumps of all sizes.
Record all of your answers. (Picture 2)

Drawn an addition square. (Picture 3)
Complete the square by adding each
vertical column number to its
horizontal row number. What do you
notice about the pattern of numbers
on your square?

+	0	1	2	3	4	5	6	7	8	9
9										
8				12				16		
7										
6										
5							12			
4										
3						8				
2			4							
1										
0					4					

What is the largest addition square of numbers that you can complete?

Subtraction on a number line

①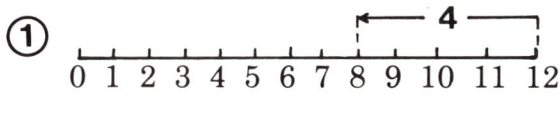

Draw a long number line (see page 12).
Addition is jumping forward (see page 12).
Subtraction is jumping backwards.

Start at 12. Jump backwards 4 spaces.
You are moving in a negative direction. (Picture 1)
On what number do you land?
Mark your jump on your number line.
We can show this jump as
    12 − 4 = 8.

Use your number line to show the jumps:
    12 − 6; 12 − 8; 12 − 7; 12 − 3; 12 − 2.

Use your number line to record backward jumps of all sizes from different starting points. Record all of your answers.

If you start at 8, this is 8 spaces from zero. If you jump backwards 5 spaces you land on 3. (Picture 2)
Notice that 3 + 5 = 8. (Picture 3)
Subtraction is the opposite of, or the inverse to, addition.
We can use this to check our answers.
    8 − 5 = 3 and 3 + 5 = 8.
Use this inverse method to check your subtraction answers.

②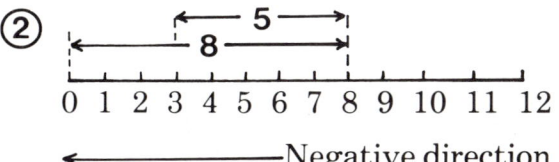

③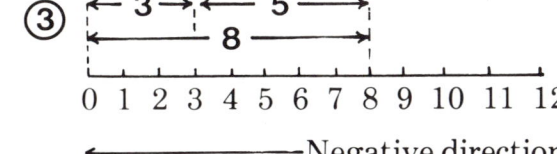

13

Addition and subtraction squares

Draw an addition square (see page 12). Complete the square by adding each vertical number to its horizontal number. (Picture 1) Look at the numbers in the square. Notice the diagonal pattern of numbers.

Look at the numbers 8 in their diagonal pattern.
Select just one square with an 8. Notice that 8 is arrived at by adding a vertical number to a horizontal number:

$3 + 5 = 8$. (Picture 1)

You can also use this square to do subtraction:

$8 - 5 = 3$ or $8 - 3 = 5$

Use your addition table to complete your own subtraction.

① Addition square

9				12						
8					12					
7						12				
6							12			
5								12		
4									12	
3						8				12
2										
1										
0										
+	0	1	2	3	4	5	6	7	8	9

Record all of your answers.

Draw a subtraction square. (Picture 2) Complete the square by subtracting the horizontal and vertical numbers. What do you notice about the number pattern that you make? Compare the addition square with the subtraction square. What do you notice?

What is the largest subtraction square of numbers that you can complete?

②

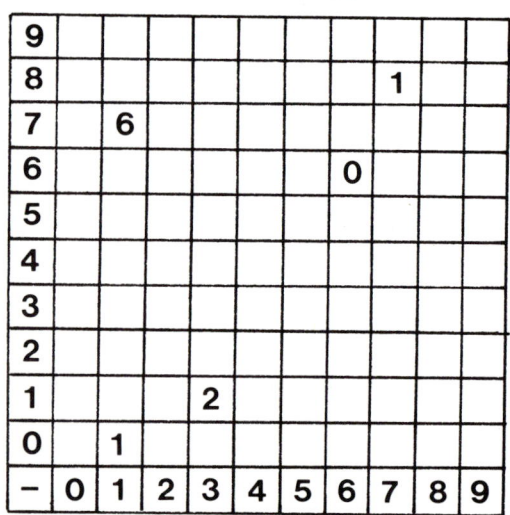

9										
8								1		
7		6								
6							0			
5										
4										
3										
2										
1			2							
0		1								
−	0	1	2	3	4	5	6	7	8	9

Subtraction square

Number sequences

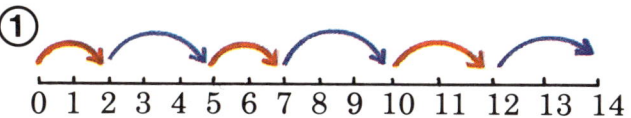

Draw a long number line (see page 12).

Imagine you own a frog that jumps in a regular pattern.

Today your frog is jumping a small jump followed by a larger jump.

Your frog's movement can be seen in Picture 1.

Your frog jumps 2 spaces, then 3 spaces, then 2 spaces, then 3 spaces and so on.

Record the numbers on which your frog lands as $0 \to 2 \to 5 \to 7 \to 10 \to 12$.

So the jumping rule is $+2, +3$.

What are the next four numbers which your frog will land on?

Sometimes your frog jumps differently.

Find the rule and the next four numbers on which your frog will land if it jumps these sequences:

$0 \to 4 \to 7 \to 11 \to 14$

$0 \to 6 \to 9 \to 15 \to 18$

$0 \to 7 \to 12 \to 19 \to 24$

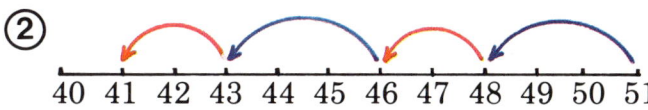

Your frog can also jump backwards. (Picture 2)

Here the jumping rule is $-3, -2$.

What are the next four numbers on which this backward jumping frog will land?

Make up your own rules for jumping frogs.

Give your friends the number sequences for each rule. Can they work out each rule and the next four numbers in the sequence?

Some frogs jump backwards and forwards. (Picture 3)

Work out some more difficult positive and negative jumps for these frogs.

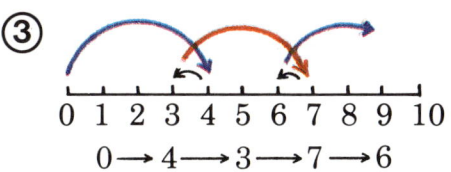

$0 \longrightarrow 4 \longrightarrow 3 \longrightarrow 7 \longrightarrow 6$

15

An integer number line

①
```
├──┼──┼──┼──┼──┼──┼──┼──┼──┼──┼──┼──┼──┼──┼──┼──┼──┼──┼──┼──┼──┼──┼──┼──┤
-11 -10 -9 -8 -7 -6 -5 -4 -3 -2 -1  0  +1 +2 +3 +4 +5 +6 +7 +8 +9 +10 +11 +12 +13
```

Draw a long number line. (Picture 1)
Put the zero in the middle of your line.
On the right of the zero put positive
numbers in order.
On the left of the zero put negative
numbers in order. (Picture 1)
You have drawn an integer number line.

Use a pencil to point to any number.
The number that you point to is
LESS than all the numbers to its right.
The sign for less than is <.
So $+8 < +12$ and $-6 < -4$.
The number that you point to is
GREATER than all the numbers to its left.
The sign for greater than is >.
So $+11 > +10$ and $-3 > -5$.
Notice that the sign always points to
the smaller number.

Select any two numbers. Write them
down and put the correct sign between them.
What sign would go between $+4$
and $+2$; $+1$ and $+3$; -2 and -4?

② +15
+10
+5
0
-5
-10
-15

The thermometer shows the
temperature 5°C below zero or -5°C. (Picture 2)
What would the temperature be if it became 3° hotter? What would the
temperature be if it was 5° hotter?
What would the temperature be if it became 7° colder?
Use the thermometer to work out your own increases and decreases in
temperature.

16

Moving along an integer number line

Draw a long integer number line (see page 16).
Start at zero and jump in a positive direction. (Picture 1) Jump 3 spaces.
You land on 3. Jump 2 more spaces in a positive direction and you land on 5.
This is addition of positive numbers (see page 12).

$$3 + 2 = 5$$

Start at —2 and jump 3 spaces in a positive direction. You land on +1.
So —2 + 3 = +1. (Picture 2)

Start at —5 and jump 3 spaces in a positive direction. You land on —2.
So —5 + 3 = —2. (Picture 3)
Use your number line to start at any number and move in a positive
direction. Record all your results.

①

You can also move in a
negative direction on your line.
This is subtraction (see page 13).
Start at +5 and move negatively 2 spaces.
You land on +3. (Picture 4)

$$+5 - 2 = +3$$

Start at +1 and move negatively
3 spaces and you land on —2. (Picture 5)

$$+1 - 3 = -2$$

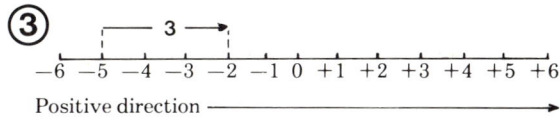

Use your number line to start at any number and move in a negative
direction. Record your results. Notice that the integer line is like a
thermometer.

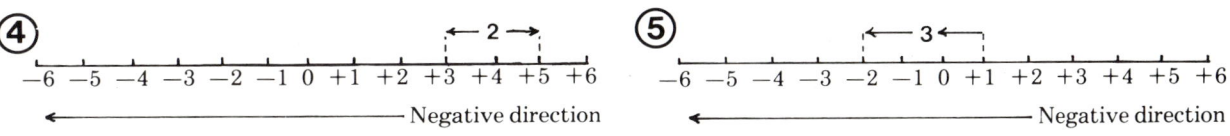

Multiplication

Draw a long number line (see page 12).
Imagine you own a frog that jumps
in regular patterns (see page 15).
Today your frog jumps only 3 spaces
each time it jumps. Starting at zero,
record all the numbers on which
your frog lands on your number line.
(Picture 1)

On what numbers will your frog land
after 4 jumps, 6 jumps, 11 jumps?
You can write this as 4 jumps of 3
spaces = 12, or 4 × 3 = 12, or 4
multiplied by 3 = 12.

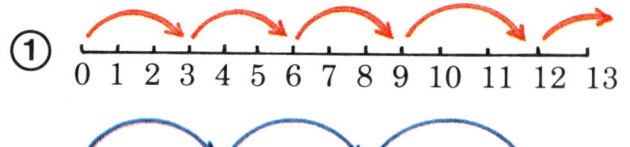

On what numbers will your frog land if each jump is 4 spaces? (Picture 2)
Notice that 3 jumps of 4 spaces = 12, or 3 × 4 = 12, or 3 multiplied by
4 =12.

Remember to always start a new series of jumps at zero.

Record the numbers on which your frog will land if each jump is 5 spaces,
6 spaces, 7 spaces, 8 spaces, 9 spaces, 10 spaces, 11 spaces and 12 spaces.

Draw a multiplication square.
(Picture 3)
Complete the square by multiplying
each vertical column number to its
horizontal row number. What do you
notice about the number patterns in
your square?

What is the largest multiplication
square of numbers that you can
complete?

③ Multiplication square

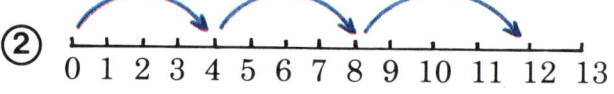

9									
8									
7								49	
6									
5									
4									
3				12					
2									
1		2							
0			0						
x	0	1	2	3	4	5	6	7	8	9

Multiplication squares

Complete your multiplication square (see page 18).
Record the numbers in the horizontal row 9.
Record the numbers in the vertical column 9.
What do you notice?

How many squares with 24 in them are there on your multiplication square? Circle each square of 24 and see which column and row it is in.
What does this tell you?
Does any number appear more often than 24?
What other numbers appear frequently?

Notice that there is only one 49 in the square.
Record all the numbers that appear only once.

Select any four squares that make a 2 by 2 square. (Picture 2)
Look at the numbers in the diagonals.
Multiply the diagonal numbers.
What pattern do you get?
Select another four squares. Do you always get the same pattern?

Write down all the numbers in one diagonal from your large multiplication square. (Picture 3)
What do you notice about this number sequence (see page 15)?
Try different diagonals. What do you notice?

① Multiplication square

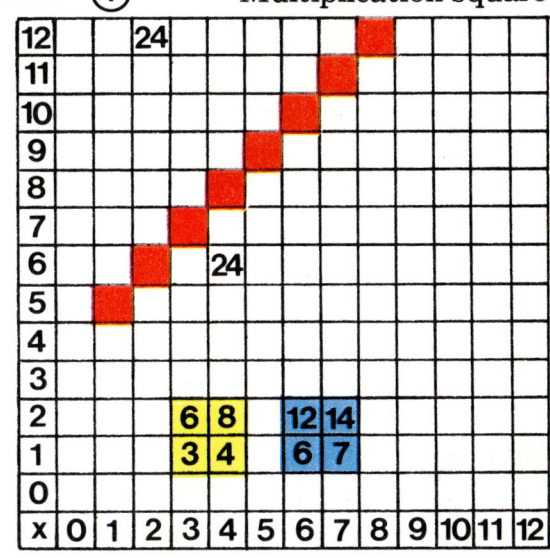

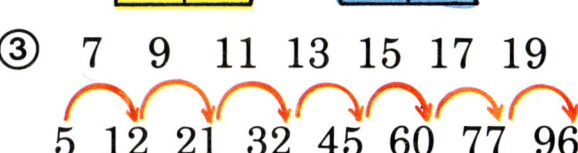

19

Division

Draw a number line (see page 12).
Imagine you own a frog that jumps in a regular pattern (see page 18).
Today your frog jumps 3 spaces in each jump. Start your frog at 18.
How many jumps does it take to reach zero? (Picture 1) We can write this as 18 spaces divided by 3 spaces = 6 jumps, or $18 \div 3 = 6$.
How many jumps of 3 spaces would your frog take to reach zero if it started at 21; 27; 36; 39; 60?

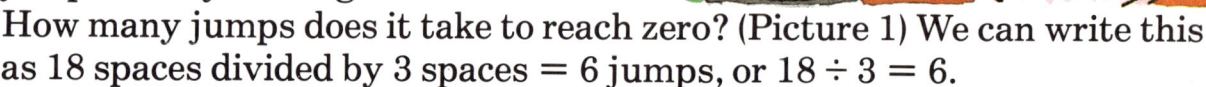

① 0 1 2 3 4 5 6 7 8 9 10 11 12 13 14 15 16 17 18

Try your frog at different numbers and with jumps of 4 spaces each.
How can you be sure that there are the correct number of jumps for the frog to land exactly on zero?

What relationship can you see between multiplication and division (see page 18)?

Division is used when you share.
If you have 24 apples and you want to share them equally between 6 friends, then 24 apples $\div$ 6 = 4 apples.
Each friend has 4 apples.

Draw a division square. (Picture 2)

② Division square

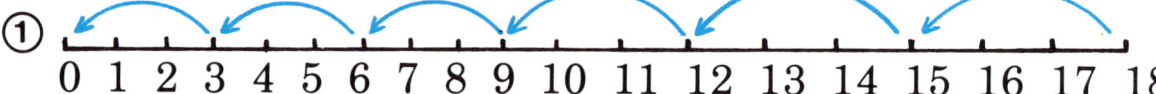

÷	0	1	2	3	4	5	6	7	8	9	10	11	12
12													
11													
10													
9										1			
8													
7													
6													
5											2		
4													
3						2							
2			1					4					
1													
0													

Complete the square by dividing the vertical column numbers by the horizontal row numbers. Only complete the squares where the numbers divide exactly.
What do you notice about the horizontal number pattern that you make?
What numbers could you fit into the blank squares to keep this pattern?
Notice these numbers are not whole numbers.

Ancient Egyptian measurements

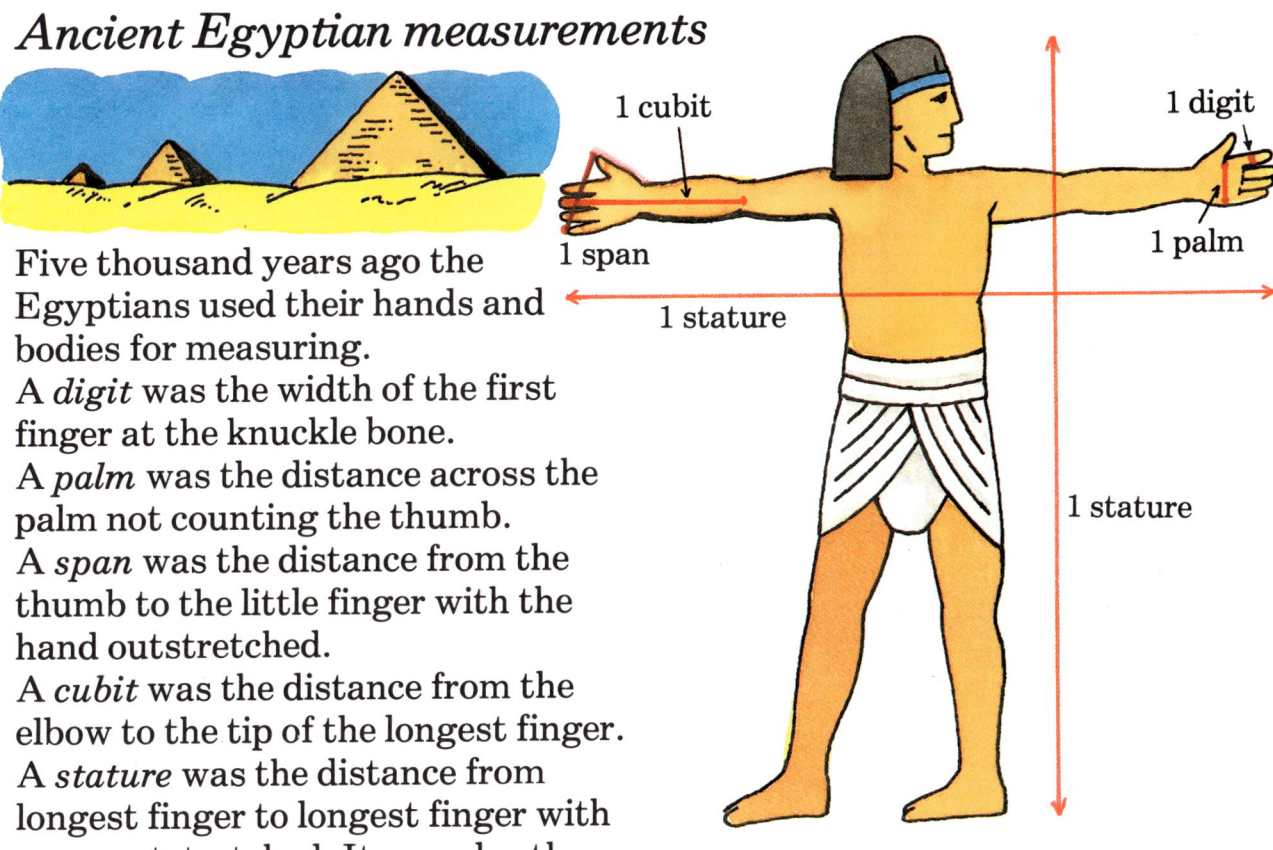

Five thousand years ago the Egyptians used their hands and bodies for measuring.

A *digit* was the width of the first finger at the knuckle bone.

A *palm* was the distance across the palm not counting the thumb.

A *span* was the distance from the thumb to the little finger with the hand outstretched.

A *cubit* was the distance from the elbow to the tip of the longest finger.

A *stature* was the distance from longest finger to longest finger with arms outstretched. It was also the height of a tall man.

Measure your digit, palm, span, cubit and stature in cm.
Record your results as a table.
Do four of your digits equal your palm?
Do two of your spans equal your cubit?

Egyptian measurement table		
1 digit		2 cm
4 digits	—1 palm	8 cm
3 palms	—1 span	24 cm
2 spans	—1 cubit	48 cm
4 cubits	—1 stature	192 cm

Repeat these measurements on your friends. Do any of them have measurements that make a table similar to the ancient Egyptian measurement table?

Do adults have measurements that match those of ancient Egyptians?

British measurements

During the reign of the English king, Henry I (1100-1135), measurements were standardized. The distance from King Henry's nose to the fingertip of his outstretched arm was called a yard.

King Henry's stature was two yards (see page 21).

They measured the king's foot and found that 1 yard was equivalent to 3 feet.

There were 12 inches in 1 foot and 4 inches in 1 hand.

In order to measure land, larger measurements were needed. (Picture 2)

How far is it from the tip of your nose to the fingertip of your outstretched hand? Do 3 of your feet measure this distance? Is your stature twice this distance? Do you know of any adult whose measurement from nose to fingertip matches King Henry's?

Did King Henry I have big feet?

② Table of British measurements

4 ins	—1 hand
12 ins	—1 foot
3 feet	—1 yard
2 yards	—1 stature
22 yards	—1 chain
10 chains	—1 furlong
8 furlongs	—1 mile
1760 yards	—1 mile

Notice that we still measure horses in hands.

To measure the sea we use different measurements from those used for land measurements.

The depth of water is measured in fathoms, which is the same as 1 stature or 6 feet.

A nautical mile is 1·15 land miles (statute miles) and a speed of 1 knot is travelling 1 nautical mile per hour.

The metric system

In 1791 the French Academy decided to have a standard length called a metre, from the Roman word for measure.
A metre was to be one ten millionth part of a quarter of the earth's circumference at sea level.
All metre rulers were made to this length.

In 1960 it was found that the French Academy had not calculated the earth's circumference accurately.
Rather than alter the length of the metre, it was decided to find something that was exactly 1 metre long.
It was decided that radiation should be used for this measurement.
A metre is divided into 100 centimetres and 1000 millimetres. A kilometre is 1000 metres.

Eight km is approximately equal to 5 miles. How many km are equal to 10 miles?
Use graph paper to draw a graph showing the relationship between km and miles. (Picture 1)
Use your graph to calculate how many km are equal to 20 miles, 30 miles, and 13 miles.

Ten inches is equivalent to 25.4 cm.
Draw a conversion graph of inches and cm. (Picture 2).

① Conversion graph km to miles

② Conversion graph inches to cm

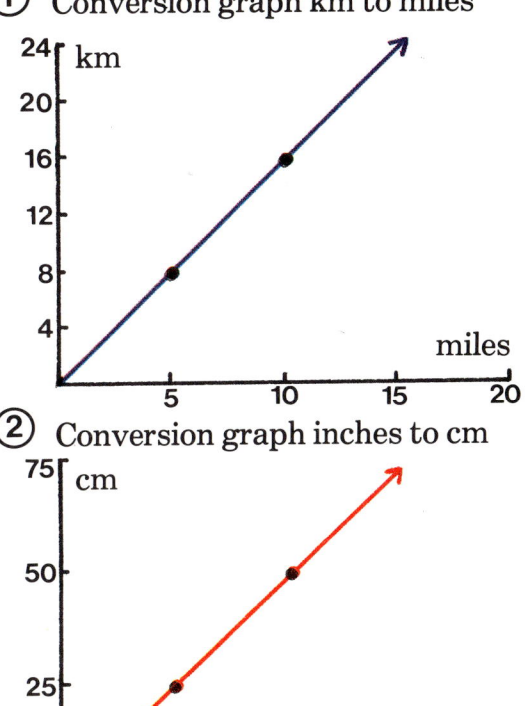

Metric measuring

Find a ruler marked in metric measurements.
Use this ruler to draw a counting board. (Picture 1)
Rule columns for metres, dm, cm, and mm.
Mark in the decimal point.

Use your ruler to draw a straight line
5 cm and 2 mm long.
Record this length on your board. (Picture 1)
Remember to mark in the zeros.
Now draw a straight line 14 cm and
5 mm long at the end of your line.
Record 14 cm and 5 mm on your board.
Add together the two lengths on your
board. Record your answer.
Measure the total length of your line.
Does this length match your answer?

Repeat this with lines of your own
length. Add together your answers on
your counting board.
Check your answers by measuring the
total length of your line.

Draw a straight line 22 cm and 7 mm long.
Record this length on your counting
board.
Mark off 8 cm and 5 mm on this line.
Record this length on your counting
board. (Picture 2)
Subtract 8 cm and 5 mm from 22 cm and
7 mm on your counting board.
What result do you get?
Measure your line to check your result.

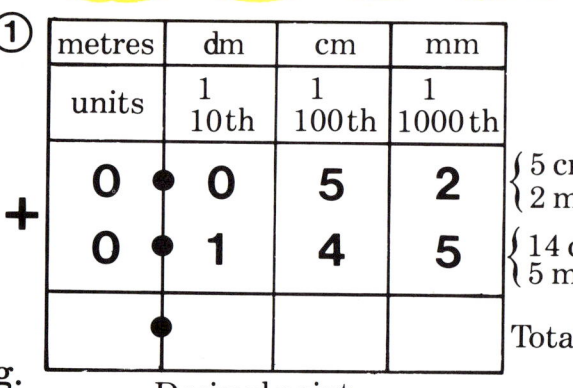

①

	metres	dm	cm	mm	
	units	$\frac{1}{10}$th	$\frac{1}{100}$th	$\frac{1}{1000}$th	
+	0 •	0	5	2	{ 5 cm { 2 mm
	0 •	1	4	5	{ 14 cm { 5 mm
	•				Total

Decimal point

②

	metres	dm	cm	mm	
	units	$\frac{1}{10}$th	$\frac{1}{100}$th	$\frac{1}{1000}$th	
−	0 •	2	2	7	{ 22 cm { 7 mm
	0 •	0	8	5	{ 8 cm { 5 mm
	•				

24

Chapter 4 Special numbers

Multiplication special numbers

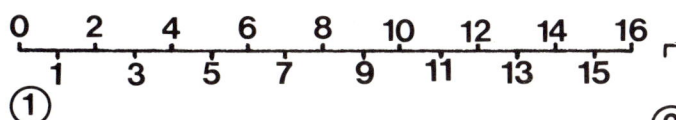

①

②

Draw a long number line. (Picture 1)
Place all the even numbers on top of the
line. Notice that if you multiply by 2 you
get even numbers. All the numbers
below the line are odd numbers.

Multiply 7, 15, 21, and 33 by 10.
Notice how easy this is (see page 10).
What do you have to do to multiply by
100; by 1000; by 10,000?

Nine is a special number for multiplication.
Number your fingers and thumbs from
1 to 10 starting on the left. Fold down
your left forefinger. The answer to 2×9
is shown by the fingers remaining.
Notice that one finger is showing on the
left and eight fingers are showing on the
right. $2 \times 9 = 18$. (Picture 2) Fold down
finger 4. Notice that you have 3 fingers
to the left and 6 to the right.

$\quad 4 \times 9 = 36$

Write down the 9 times table. Look at
the numbers. Why does this finger trick
work? Here are some more
multiplications by 9. (Pictures 3, 4, 5)
Try to explain these results.

③
$$999999 \times 2 = 1999998$$
$$999999 \times 3 = 2999997$$
$$999999 \times 4 = 3999996$$
$$999999 \times 5 = 4999995$$
$$999999 \times 6 = 5999994$$
$$999999 \times 7 = 6999993$$
$$999999 \times 8 = 7999992$$
$$999999 \times 9 = 8999991$$

④
$$123456789 \times 9 = 111111111$$
$$123456789 \times 18 = 222222222$$
$$123456789 \times 27 = 333333333$$
$$123456789 \times 36 = 444444444$$
$$123456789 \times 45 = 555555555$$
$$123456789 \times 54 = 666666666$$
$$123456789 \times 63 = 777777777$$
$$123456789 \times 72 = 888888888$$
$$123456789 \times 81 = 999999999$$

⑤
$$9 \times 9 + 7 = 88$$
$$98 \times 9 + 6 = 888$$
$$987 \times 9 + 5 = 8888$$
$$9876 \times 9 + 4 = 8888$$
$$98765 \times 9 + 3 = 88888$$
$$987654 \times 9 + 2 = 888888$$
$$9876543 \times 9 + 1 = 8888888$$
$$98765432 \times 9 + 0 = 88888888$$

Prime numbers

A prime number cannot be divided by any number except 1 and itself. One is not a prime number as 1 is itself. Two is a prime number, as is 3. Four is not a prime number as 4 can be divided by 2.

Write down in order all the whole numbers from 2 to a 100. Circle all the prime numbers. What do you notice about the prime number pattern that you make (see page 15)?

Eratosthenes, an ancient Greek, discovered a way to find prime numbers. Write down in order all the whole numbers from 2 to a 100. (Picture 1) Circle the first prime number 2. Now cross out every SECOND number after this. The first number left is 3. Circle the 3. Now cross out every THIRD number after this. The first number left is 5. Circle the 5. Now cross out every FIFTH number after this. Circle the 7. Then cross out every SEVENTH number after this. The next prime number to circle is 11.

What do you think Eratosthenes did now? What is the largest number of prime numbers that you can find?

It is said that EVERY even number bigger than 6 can be written as the sum of two prime numbers. Try it and see. (Picture 2)

① ②③ ✕ ⑤ ✕ ⑦ ✕ ✕ ✕ ⑪ ✕ ⑬ ✕ ✕ ✕ ⑰ ✕ 19 20 21
22 23 24 25 26 27 28 29 30 31 32 33 34 35 36 37 38 39 40 41 42

②   5 + 7 = 12
      7 + 13 = 20

Numbers by shape

Find some graph or squared paper.
Draw perfect squares on the lines
of your paper.
Make your perfect squares different
sizes. (Picture 1)
Count the number of graph paper
squares there are within the perimeter of
your drawn square. Record your results.
How many graph paper squares are
there along only one side of each square?
Notice you can make a square of 4
squares, 9 squares and 16 squares.

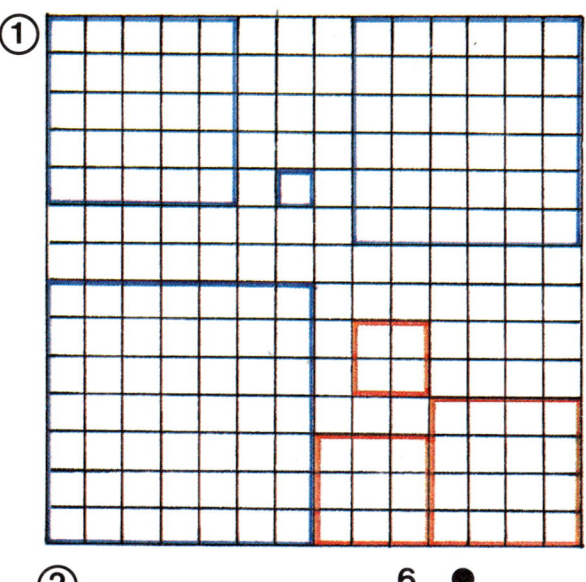

These are square numbers.

$2 \times 2 = 4; 3 \times 3 = 9; 4 \times 4 = 16$

Notice the pattern. What is the next
square number? How many square
numbers can you make?

Find some equal sized coins or washers.
Arrange your coins into triangle shapes.
(Picture 2)
Record the number of coins in each
shape.
How many coins are there along one side
of each triangle?
Notice that to make larger triangles you
add a bottom line that has one extra coin
in it.
How many triangle numbers can you
calculate?

27

Index numbers

To square a number you multiply that number by itself (see page 27).
Three squared is nine or $3 \times 3 = 9$.
We can show this in two dimensions by writing $3^2 = 9$. The 2 is the index and means 3 to the SECOND power equals 9.

A cube is in three dimensions. It has length, width and height.
It takes 27 small cubes to make one cube with faces 3×3. (Picture 1)
$\quad 3 \times 3 \times 3 = 27$ or $3^3 = 27$
Three to the THIRD power is 27 or 3 cubed is 27.

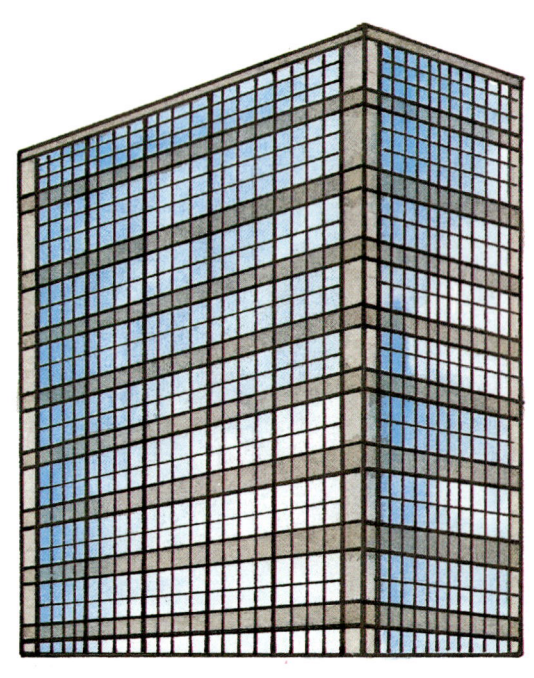

①

②

Index counting board		
2^0	–	1
2^1	2	2
2^2	2 x 2	4
2^3	2 x 2 x 2	8
2^4		
2^5		
2^6		

What is 2 cubed? What is 4^3?

We cannot draw the fourth dimension, but we can calculate numbers in it.
$\quad 2 \times 2 \times 2 \times 2 = 16$ or $2^4 = 16$
What is $3 \times 3 \times 3 \times 3$?

Draw the index counting board. (Picture 2)
Complete the counting board.

Notice that the index 1 is the number in one dimension or just the number.
The index zero always gives the answer 1.

Draw index counting boards for 3, 4 and 5.
What is the highest power that you can calculate?

Number columns

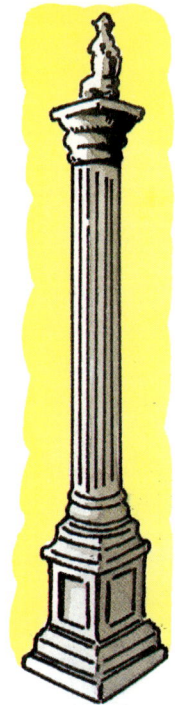

Draw three columns. Put whole numbers in each column in order. (Picture 1)
Label your columns A, B and C.
Add together any two numbers in column A.
Record the column in which you can find the answer. Is it always the same?

Add together any number from column A with any number from column B. In which column can you find the answer?
Add together any number from column B with any number from column C. In which column do you find the answer?
Try columns C + C; A + C; B + B.
What pattern do you get?

Draw five columns. Put whole numbers in each column in order. (Picture 2) Label your columns A, B, C, D and E.

Add any number from column C to any number from column D. In which column do you find the answer? Is it always the same? Add together any number from any column to another number from any column. Record your answer as a table. (Picture 3)

Does this work if you have numbers in seven columns?
What do you know about the number of columns needed for this to work?

①

A	B	C
0	1	2
3	4	5
6	7	8
9	10	11
12	13	14
15	16	17
18	19	20
21	22	23
24	25	26

②

A	B	C	D	E
0	1	2	3	4
5	6	7	8	9
10	11	12	13	14
15	16	17	18	19
20	21	22	23	24
25	26	27	28	29
30	31	32	33	34
35	36	37	38	39
40	41	42	43	44

③

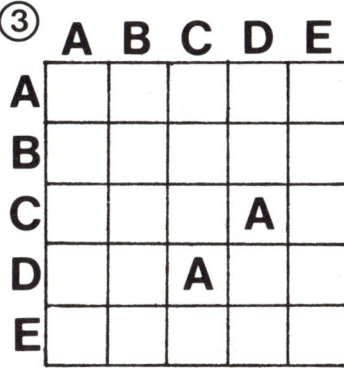

	A	B	C	D	E
A					
B					
C				A	
D			A		
E					

Fibonacci numbers

(1) Fibonacci numbers

0
1
1
2
3
5
8
13
21
34
55
89
144
233
377
610

In the year 1202 Leonardo Fibonacci of Pisa, Italy, created a number sequence. (Picture 1)

Each number in the series is made by adding together the two previous numbers.

$$0+1=1; 1+1=2; 1+2=3; 2+3=5$$

How many more numbers can you add to Fibonacci's sequence?

Draw a line under any number in the column. The sum of all the numbers above the line is equal to one less than the second number below the line.

Try it for yourself. Put a line under 21. What is $0+1+1+2+3+5+8+13+21$? The second number below the line is 55.

$$55-1=54$$

Take any three numbers in sequence from the Fibonacci numbers. Square the middle number (see page 27). Multiply the first and third numbers together. The difference between the two answers is always the same. (Picture 2)

Many natural things follow the Fibonacci sequence. Find a flower stalk. If the lowest leaf is zero, count the leaves on the stalk until you find one directly above your starting leaf. Record this number.

Notice that this number is a Fibonacci number.

(2) 3, 5, 8

$$5 \times 5 = 25$$
$$3 \times 8 = 24$$

Difference $25-24=1$

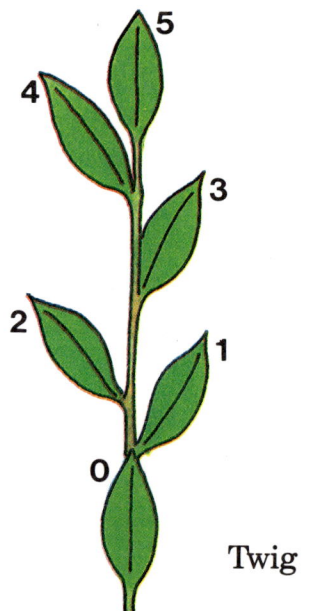

Twig

30

Binary numbers

Number the fingers and thumb of your right hand as in Picture 1. You can show all the whole numbers from zero to 31 with this hand. To make a number, hold up the fingers needed to add up to that number.

For 1 you only raise your thumb.

For 2 just raise the next finger. For 3 you hold up both of these fingers. (Picture 2)

What finger do you hold up for the number 8?

How can you show the number 9?

Use your hand to show all the numbers to 31.

To make numbers up to 1023 number the fingers of your left hand as in Picture 3.

Show your friend this code and use fingers to show different numbers.

Go back to your right hand.

You can write this code by agreeing that a finger up is 1 and a finger down is 0.

Now 21 can be written as 10101. (Picture 4)

What number can be written 10111?

Write your own numbers in this code.

This code is the BINARY code as it uses only the numerals 0 and 1.

Electronic digital computers use the Binary system. When the current is flowing through a circuit, that is 1. When the current is not flowing through a circuit, that is 0.

Have you used a computer?

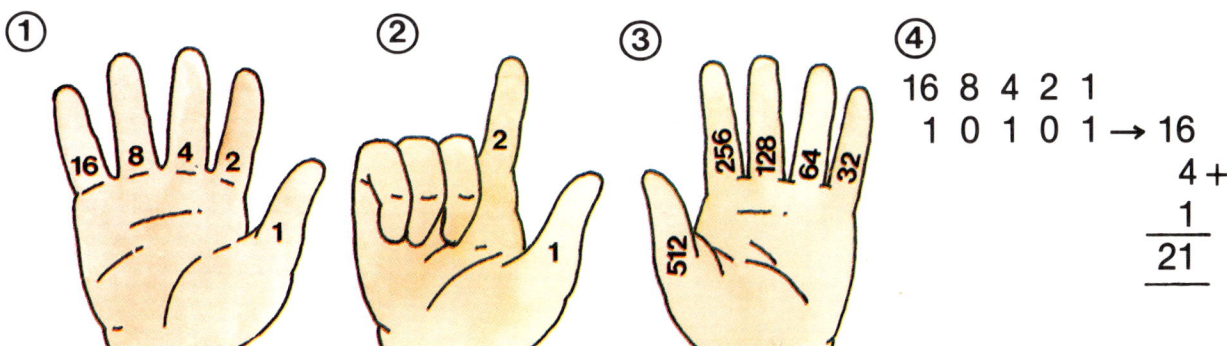

④
16 8 4 2 1
1 0 1 0 1 → 16
         4 +
         1
         ――
         21

Number bases

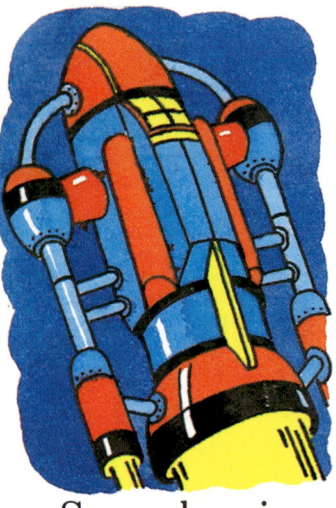

We have ten fingers on our hands.
We use ten symbols for our numbers (see page 9).
We use these ten symbols in a place value code (see page 10).
Sixty-three is six 10s and three 1s.
Our value code is in 10s. (Picture 1)
Our number code is 10 or we can say we work in base 10.

You can also use a binary code (see page 31).
Here you work in base 2. (Picture 2)
110 base 2 is 4 plus 2 which makes 6.
What is 11011 base 2?

Somewhere in space there may be other creatures that can count. They may have six legs like an insect or eight legs like a spider. They may use a different base code.

Make a place value counting board for base 3. (Picture 3)
120 base 3 is one 9 and two 3s which is 15.
Put your own numerals on your base 3 counting board. Calculate what your numbers mean.
Remember 2 is the largest number that you can put in any base 3 column.

Make a base 4 counting board. (Picture 4)
Put your own numerals on this board.
Calculate what your numbers mean.
The columns for base 5 are 1, 5, 25, 125 and 625.
Calculate in base 5.
What would be the columns for base 6, base 7, and base 8?
What is the largest number that you can put in any column in these bases?

① Base 10

100s	10s	1s	
	4	5	45
6	5	3	653

45 is four 10s and five 1s
653 is six 100s and five 10s and three 1s

② Base 2

16	8	4	2	1	
			1	1	3
		1	1	0	6
1		1	1		22

③ Base 3

81	27	9	3	1	
		1	0	1	10
	1	0	2	0	33
	1	2	2	2	53

④ Base 4

256	64	16	4	1	
		1	2	3	27
	1	2	3	3	111

32